People and Progress

Coal Mining

Don Hale and Martyn Vickers

Miners on strike, 1912

1 The Lofthouse Disaster

Daily Mirror, 21 March 1973

TRAPPED

Hopes fade for the seven miners in a flooded hell-pit

In the night of 21 March, 1973, 15 miners were working at a **coal face** 250 metres under the ground at Lofthouse Colliery in West Yorkshire. Suddenly, they heard a great bang 'like a big crack'. Then they watched in horror as the wall crumbled in front of them. Millions of litres of water poured in on them with such force that it knocked men down.

THEIR 1,000 METRE DASH FROM DEATH

All the miners could do was to run but it was over 1000 metres to where the mine passages sloped upwards. So they desperately raced from death as the water chased them down the passage-ways. Only the youngest and fastest kept ahead of the water and escaped the flood.

Eight men reached safety but seven were trapped by the muddy water. It was just possible that the trapped men might have reached an air pocket in a higher part of the mine and so a desperate rescue was started to try to reach them.

The rescuers did not have much time to find the trapped men. Experts said that the men could only have about 40 hours of air left. Giant pumps were set up. **Shafts** were dug down and rescuers and frogmen worked day and night to try to reach the men. An enormous wall of mud held up the rescue and time began to run out.

A rescue team at the Lofthouse pit

ONLY A MOUSE WAS ALIVE..

The scene inside the Lofthouse pit

On 26 March the rescuers finally reached the place where they hoped the men would be. There was nobody there: only a mouse was alive. The waters and the mud had claimed the seven men. One body was found but those of the other six could not be reached. Coal miners would not forget the name 'Lofthouse'.

The first sign of what had caused the disaster at Lofthouse came when a man who was out walking in a field near to the mine saw a gaping hole about 3 metres wide appear suddenly in front of him. Soon two more holes appeared. All three holes were the shafts of old coal mines which had been worked about a hundred years before until they became flooded and were filled in. By mistake, the Lofthouse miners had cut into a flooded mine.

The National Coal Board knew about the old mines but they did not know that the maps were wrong. The old mines went much deeper than the maps showed. An old notebook was discovered which told the truth. Better maps a hundred years ago could have saved the miners' lives. Coal mining has always been dangerous and you can find out more about the problems of mining through the ages in this book.

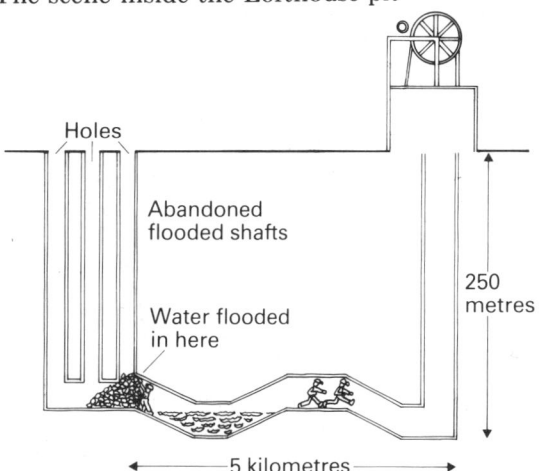

The Lofthouse colliery

Things to do

1 You are a newspaper reporter sent to Lofthouse Colliery on 21 March 1973. Write a report to say what happened that day.
2 Write a few lines to say how the Lofthouse disaster happened. Use these clues to help you: old mines; flooded workings.
3 'Better maps could have saved the Lofthouse miners' lives'. Why?
4 Start a scrapbook of newspaper cuttings about the coal industry today.
5 Is there a coal mine near to where you live? If so, see what you can find out about it. If not, see if you can find out where in the country coal in your town comes from.

2 Early Coal Mining

In very early times, hundreds of years ago, coal was gathered on the beaches and shores where the sea washed it up from coastal **outcrops**. Later on, men began to dig the coal from these **seams** to use as a fuel in place of wood. Coal mining had begun.

Gathering coal on the sea shore

A drift mine

As man dug out the coal seams he made the first coal mines known as **drift** mines. Later, shafts were dug into the ground where coal was known to be. Some of these mines were called 'Bell Pits' because of their shape.

In these the coal was dug out with simple tools until the **pit** became too dangerous to work. The two main dangers were from flooding as water seeped in through cracks in the rock and the threat of rock falls as the miner dug out the coal. When the pit became too dangerous it would be abandoned and a new one begun. This would often be at the spot where the miner's shovel landed after he threw it from the old pit.

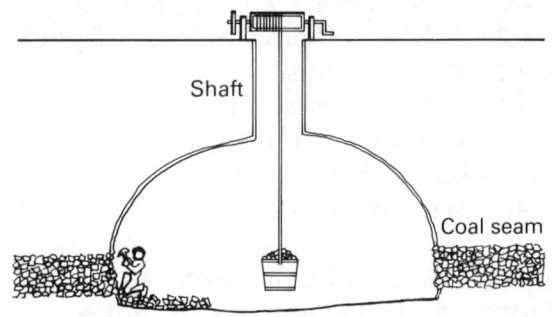

A bell pit

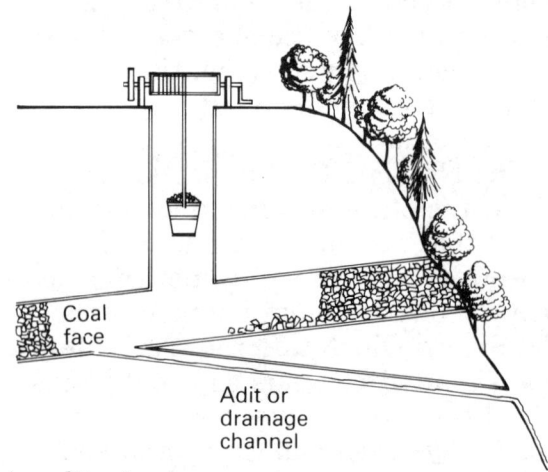

An adit mine

Other early pits were known as **adit** mines. An adit was a drainage channel dug from the **coal face** so that the water would flow away from where the men were working. Most of these early mines were very small, not usually going deeper than about 10 metres.

Coal mining in those early days was simple but even then coal was being mined in **coal fields** all over Britain.

This map shows you the coal fields of Britain today. Coal was being mined in nearly all these coal fields over 200 years ago.

One big problem at that time was that the roads were so bad that coal could not be sent anywhere by land. So the most important coalfield was the one in Northumberland because it was near the sea.

Barges called keels took the coal down the rivers to the sea where it was loaded on to ships called **colliers**. These ships were used to carry the coal from Northumberland to London so that people there could burn coal in their fires.

This coal was usually known as 'sea coal' because it came by ship. In those days only people who could not afford wood burned coal in their fires. As the years went on though, wood became scarce and more expensive so more and more people began to have coal fires.

The coalfields of Britain

A Keel

A Collier

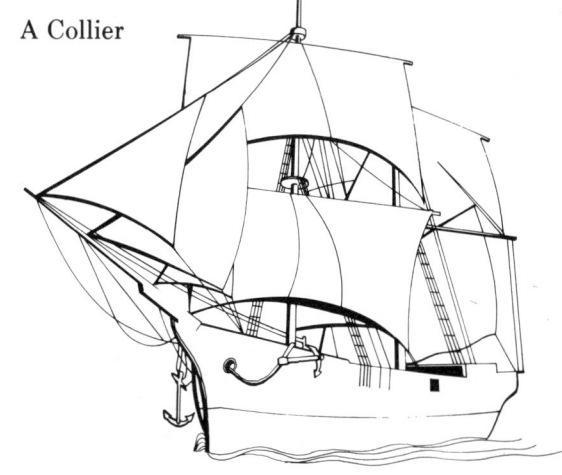

Things to do

1 Make your own drawing of a bell pit. Write two sentences to describe two dangers which bell-pitmen might have had to face.
2 Write a sentence on each of the following, explaining their meaning: coal seam, outcrop, shaft, coal face, adit, coal field.
3 Draw a collier (ship). Find out what else this word can mean.
4 Find out the ways by which coal is carried today.

3 Going Deeper

Coal mining was to change rapidly in the years after about 1700.

About that time Abraham Darby found that he could make iron by **smelting** (or burning) iron ore with **coke**. For the next 150 years iron makers needed more and more iron to make guns, rails, bridges and many other things. Wood was so scarce that things once made out of wood now came to be made of iron. This was part of an Industrial Revolution which was changing Britain. Big new factories were being built for new machines

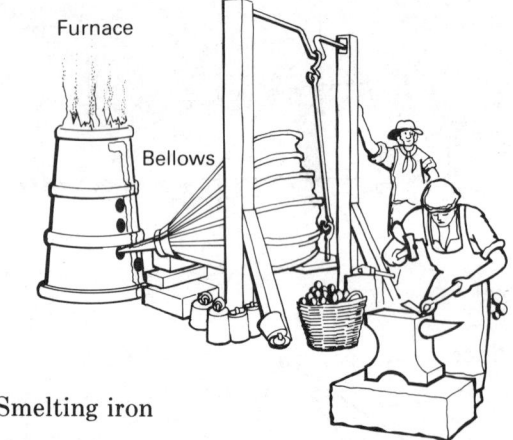

Smelting iron

driven by a new kind of power—steam power. These new steam engines burned coal as their fuel.

So coal was needed more to make iron and to feed the new steam engines. Soon coal supplies near the surface ran out and miners had to go deeper to get the precious fuel.

These diagrams show you how mines have had to be dug deeper as more coal was needed.

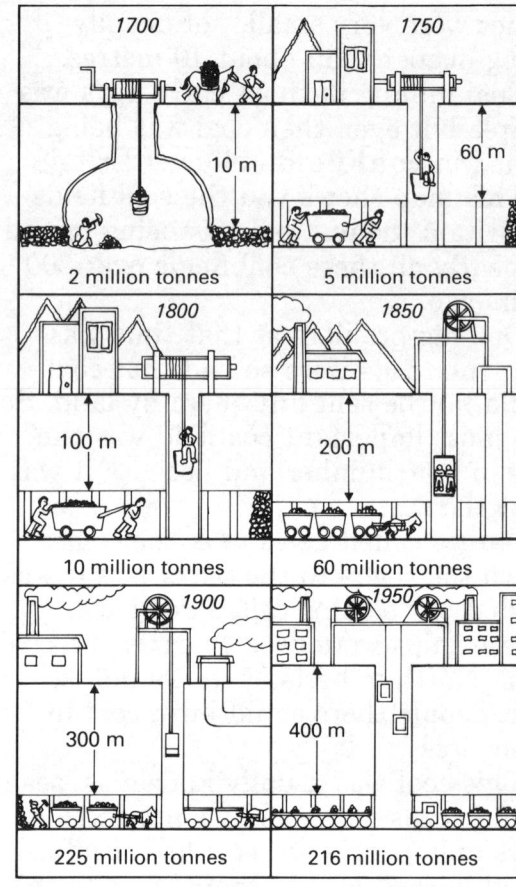

1700	1750
10 m	60 m
2 million tonnes	5 million tonnes
1800	1850
100 m	200 m
10 million tonnes	60 million tonnes
1900	1950
300 m	400 m
225 million tonnes	216 million tonnes

Going deeper to dig more coal

As the mines became deeper, coal mining became more dangerous. New hazards were found and many serious problems had to be solved.

Things to do

1 Make your own graph to show how coal production increased in these years: 1700, 1750, 1800, 1850, 1900.
2 Why did iron making and steam engines mean that Britain needed more coal?
3 Abraham Darby, Henry Cort and John Wilkinson are very important in the history of the iron industry. Find out why.

4 Stale Air

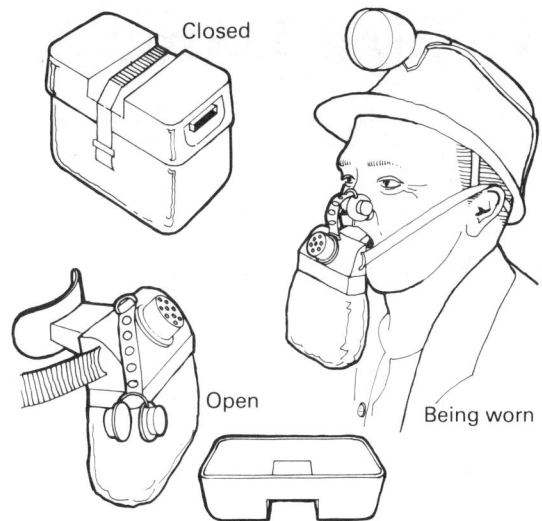

Closed

Open

Being worn

A modern self-rescuer

If the air that we breathe is not changed regularly, it becomes stale and unhealthy. As the miners dug deeper the air quickly became stale. Gases coming out of the coal made the air even more dangerous to breathe.

One of the gases, carbon dioxide, choked and suffocated the miners so they called it choke damp. By watching the flame of his candle grow dim the miner would know that choke damp was present and he would **waft** away the gas with his jacket or a rag. If a man was badly affected by choke damp he would be brought to the surface. His face would be forced into a newly dug hole in the ground and soil packed round it to try to force his lungs to cough out the gas. Although choke damp is dangerous, a far more deadly gas was left in the mine after a fire or explosion. This was carbon monoxide or after damp. It is very poisonous and gives little warning. A man affected by the gas feels giddy or sleepy and unless he quickly gets fresh air he will die.

For many years canaries were taken down mines as they are affected by the gas before people and so gave miners an early warning. In the Gresford Mine disaster in 1934 which killed 265 men one of the rescuers noticed that 'the canary I was carrying was dead and so we could not go in any further.'

Today, chemicals which change colour in gas are used to detect poisonous gases in most mines but canaries are still used at times. Also, every miner must take a self-rescuer down the mine with him to protect him when gas is about. By breathing through the mask the poisonous gases are changed into harmless ones. The self-rescuer will only protect the miner for about one hour.

The best way to protect miners from these gases is to ventilate the mine by changing the air so that no stale air or dangerous gases remain in the mine.

> **Things to do**
> 1 Make a table with four columns about poisonous gases found in mines. In the first column put the miners' names for the gases. In the second put the real names of the gases. In the third write down how the gases affect people. In the last write down how we deal with each gas.
> 2 How could a miner tell if Choke Damp was around? What would he do about it?
> 3 You have to show a miner how to use a self-rescuer. Explain how it works and why it is used.

5 Fire Damp

A warning notice

As the miners dug deeper still they met the most dangerous gas of all—fire damp or methane. Although it is not poisonous, it is highly explosive and easily catches fire. A miner's candle would make the gas explode without warning and the results were terrible. In the 18th century there were many disasters caused by fire damp. Here are details of one in the North-east. *1708: Wear Colliery:* '69 were killed. The blast blew the bodies of two men and a woman 110 metres up the shaft and out of the pit. The bodies were terribly mangled. The explosion was like a thunder clap and the shock waves killed fish in local rivers.'

Methods of trying to deal with fire damp were very dangerous. Some pits had a highly paid 'fireman' who dressed in wet rags and crawled into the gas with a candle on the end of a long pole. When the flame reached the gas it caused an explosion. The 'fireman' took his chance and ducked.

Sometimes candles were drawn into the mine on trolleys pulled by ropes so that the man pulling the rope could be at a safe distance. In other pits, ponies were driven into the gas with candles on their backs—killed to keep the mines open. All these ways were clumsy and dangerous. Miners risked their lives by working in gas.

The main problem was that miners needed light to work by and it was the flame from their candles that set fire to the gas. No wonder that some men worked in the dark while others refused to go down gassy mines.

Things to do

1 Make three drawings to show early methods of dealing with fire damp.
2 You are taking a miner down a mine for the first time. Explain to him why it is important that he is searched before going underground.

An explosion caused by fire damp

A 'fireman'

6 Light and safety

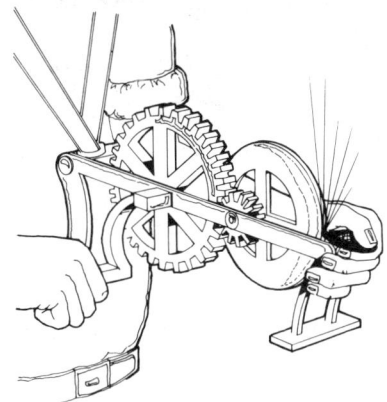

Spedding's steel mill

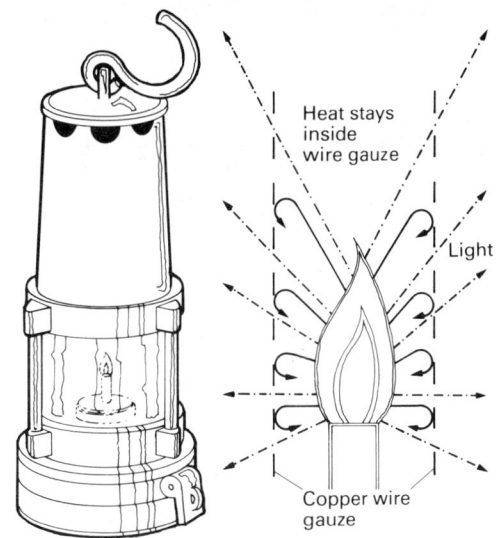

The Davy lamp and how it works

Miners tried many ways to try to find a light which would not set fire to the gas. Carlyle Spedding, who was later to be killed in a fire damp explosion, invented a 'steel mill' which made a shower of sparks. The sparks gave only a glimmer of light and the 'mill' needed an extra person to use it.

Meanwhile the accidents went on and more and more people became convinced that something had to be done. After an explosion in 1812 at the Felling Colliery, the local minister, Rev. John Hodgson formed the Society for the Prevention of Accidents in Coal Mines. This Society asked the great scientist, Sir Humphrey Davy to look into the problem of making a safe light for mines. Davy invented his Safety Lamp in 1815 just as another inventor, George Stephenson, made his own lamp. It was an instant success and was soon being used in many mines.

Davy's lamp was very simple. He surrounded the lamp flame with a wire gauze. This let air and some gas in but it did not let the heat of the flame out. However much gas got into the flame, the gas outside the gauze never got hot and so did not explode. It gave a safe light and it served another purpose as the small flame gave the miners warning when fire damp was about.

Davy's great invention has hardly been altered to this day and it is still used to test for gas. Strangely for many years it did not reduce accidents in the mines. This was because miners were now sent to work in very gassy and dangerous parts of the mines where they could not work before. Other accidents were caused by stupidity—like taking the gauze from the flame to light a pipe. Eventually the use of electric lights gave an even safer light, but even now strict checks must be made to make sure that miners do not take anything down a mine which might cause a fire or an explosion.

Things to do

1 Make a drawing of Spedding's steel mill. Write down one good and one bad thing about the invention.
2 Make a poster for the Society for the Prevention of Accidents in Mines advertising for someone to invent a safety lamp for the mines.

7 Ventilation

Using bellows to ventilate a mine

Ventilation of a mine means taking stale or impure air out and replacing it with fresh air. If a mine is to be properly ventilated, fresh air must move around all parts at all times. There are four important reasons why mines need to be ventilated.

1 To bring in oxygen to allow miners to breathe and work underground.
2 To take out dangerous gases such as choke damp and fire damp.
3 To carry away dangerous coal dust which can cause explosions and serious illness.
4 To cool the working areas.

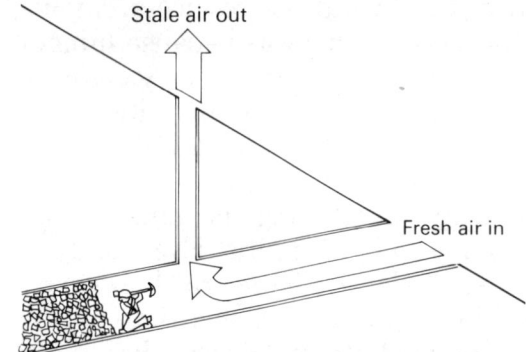
Using a shaft to ventilate a mine

By digging a shaft from the mine to the surface in a drift mine, a flow of air could be made. The air, warmed slightly by the miner's bodies and their candles, would rise naturally up the shaft and be replaced by cool, fresh air drawn into the mine.

Huge bellows were sometimes used to blow fresh air down the shaft through wooden pipes to where the coal was mined.

About 300 years ago, the first pits with two shafts were sunk. This helped because there was usually some natural ventilation. Miners soon realized they could improve the flow of air in the mines by warming the air with fire.

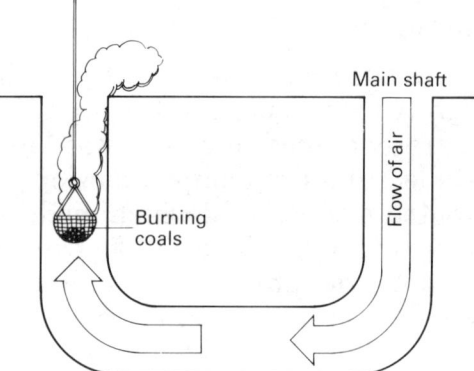
Circulating the air with burning coals

Sometimes a basket of burning coals would be lowered down one shaft drawing stale air out and allowing fresh air in.

An 'air coursing' system

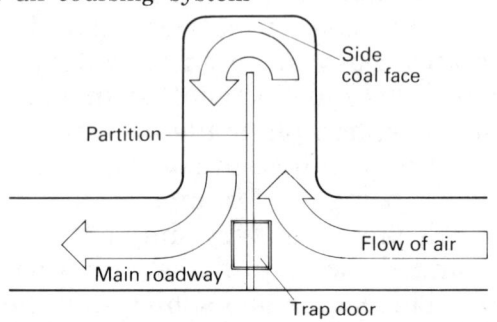

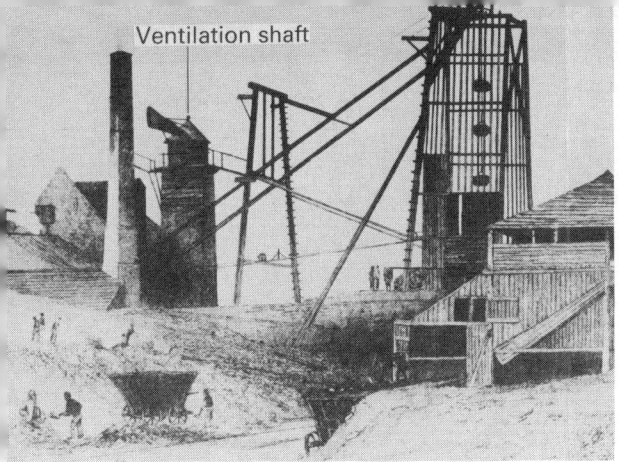

Ventilation shaft

A ventilation system at the pit head

Using fire was very dangerous and many serious explosions occurred. About 1760 a new method of ventilation was invented called 'air coursing' which made the fresh air travel along all the underground roads before leaving the mine. This meant that there was less chance of small **pockets** of gas building up. But it could mean that the air had to travel a long way (up to 50 km in some pits) before leaving the mine. By that time it was possible for it to contain large amounts of dangerous gas or dust.

By 1810, a new idea was being tried for the first time by John Buddle. As the air entered the mine it was split and by using trap doors often worked by children, was sent to the different coal faces and workings. This helped a good deal and was quickly copied in many other pits. But the main problem still had not been solved—fire was still being used and explosions continued to happen.

This verse from a popular folk song tells the fears of miners who worked in poorly ventilated pits.

Down shafts ill-ventilated the miner he must go
And crawl upon his hands and knees where'er the roof is low
The miner, fireman, driver and the trapper in his hole
Are all exposed to danger whilst down among the coal.

John Buddle was the first to use a ventilation system that did not use fire to move the air round the mine. In 1807 he used the first mechanical ventilator at Hebburn Colliery which pumped stale air out of the mine at the top of one of the shafts.

During the next 50 years this idea was developed and improved upon and in 1849 William Brunton used the first successful fan to ventilate mines.

Today it is illegal to use fire to ventilate mines. Now most mines are ventilated by large electric extractor fans placed at the top of the shafts.

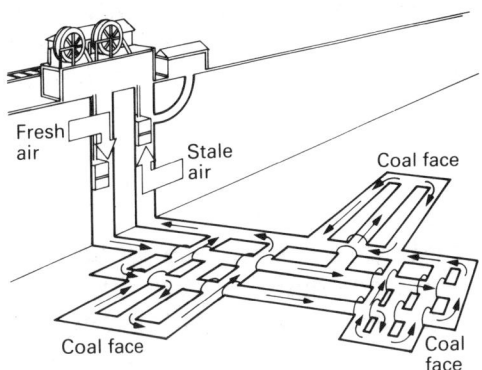

Fresh air Stale air Coal face Coal face Coal face

A modern ventilation system

Things to do

1 Make four pictures, similar to those in a strip cartoon, to show why proper ventilation is so important in coal mines. Write a caption for each picture.
2 Make your own drawings of three early methods of ventilating mines.
3 Write a conversation between two coal mine-owners. One is pointing out the advantages of using two shafts to ventilate a mine using a furnace at the shaft bottom. The other points out the dangers and disadvantages of the system.

11

8 Water in Mines

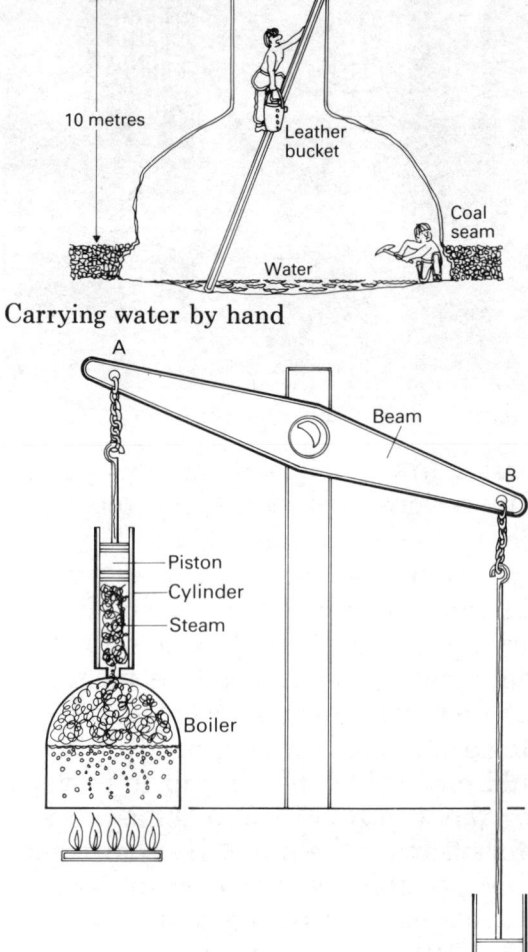

Carrying water by hand

Newcomen's engine

Another serious problem facing miners in the first coal mines was water. The water seeped through the rock above where the miner worked to flood the mine. As mines went deeper, the problem of water got worse. A tragic accident caused by a violent thunderstorm took place in 1838 at Silkstone in South Yorkshire. Twenty-six young people aged between 7 and 17 were drowned—trapped against a ventilation door by a sudden flood of water in the mine.

At first, the only way to get rid of the water was by scooping it up into buckets made of wood or leather to be carried up to the surface by hand.

Drainage channels were dug later, sometimes up to two miles long, to drain away the water. By the 1500s simple machines were being used. A popular one was an endless chain of buckets on a rope wound up the shaft by horse, water or man power.

As the pits went deeper, better ways of raising the water were needed. In 1698 Captain Thomas Savery invented a steam pump which he called the 'Miner's Friend'. Unfortunately, it could only raise water about 10 metres at a time and used enormous amounts of coal, and so was not very useful.

About 1712 a far better method was invented by Thomas Newcomen. This engine moved rather like a see-saw. Steam was let into the cylinder from the boiler. When the cylinder was full of steam a spray of cold water changed

the steam back to water (that is, it condensed the steam). This helped to make the piston fall and pulled end **A** of the beam down to work the pump. Then a weight pulled the end of the beam at **B** back down so that the movement could start again. Newcomen's engine could pump water from depths of over 100 metres and used much less coal than the 'Miner's Friend'. It became very popular and was used in many mines all over the country for more than 60 years.

A Scotsman, James Watt, was repairing a model of one of Newcomen's engines when he saw how he might improve it. He did this by using a separate condenser (the condenser changes the steam back to water to make the beam move). He was helped by Matthew Boulton who gave Watt money to make more improvements. He worked on the problem of getting the steam engine to drive a wheel, that is, to give a rotary motion. In 1781 he succeeded in doing this with his 'Sun and Planet' gear. These and other improvements made Watt's engine much better and cheaper than the Newcomen pump.

By 1850 most pumps were placed at the bottom of the shaft and after 1881 powerful electric pumps were used at many pits.

Despite these improvements mines can still be miserably wet as they were in the 19th century when a visitor to a pit wrote: 'There was a continual drip of water and the miners worked with nothing on except a pair of ragged trousers and clogs.'

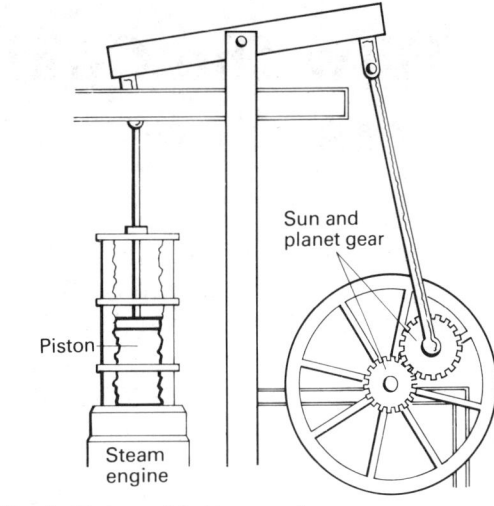

Sun and planet gear
Piston
Steam engine
Watt's Rotary Motion engine

Things to do

1 Make an advert to sell a James Watt steam engine 200 years ago.
2 Explain how a Newcomen pump works to a mine owner in 1730.

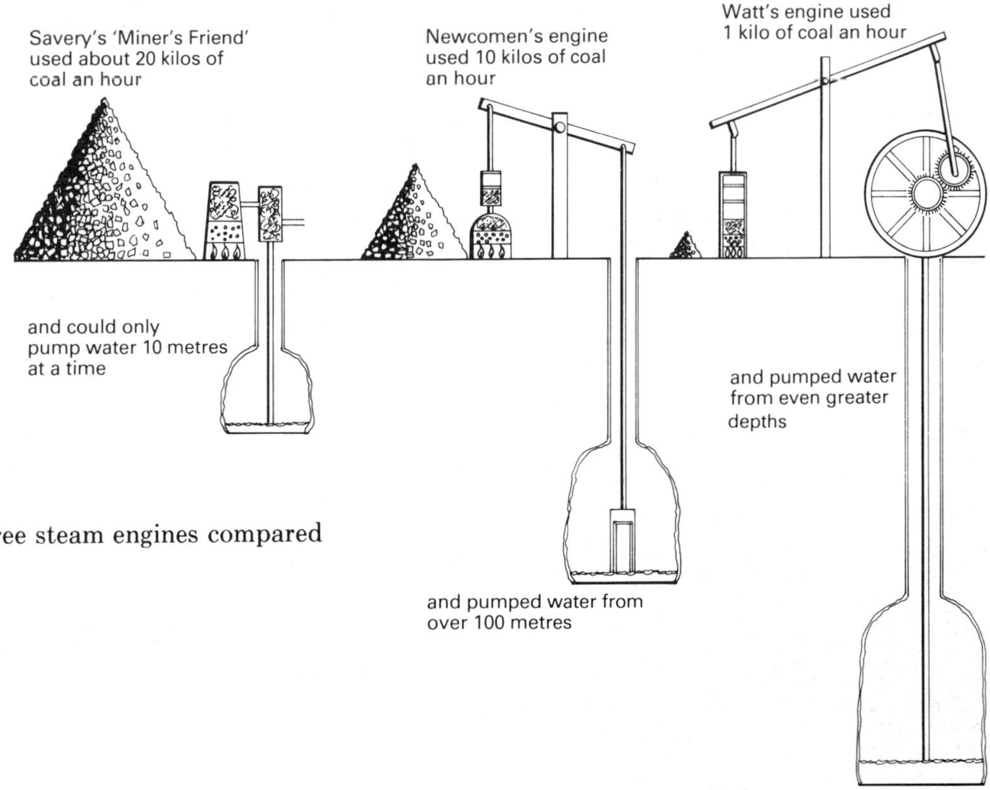

Savery's 'Miner's Friend' used about 20 kilos of coal an hour
and could only pump water 10 metres at a time
Newcomen's engine used 10 kilos of coal an hour
and pumped water from over 100 metres
Watt's engine used 1 kilo of coal an hour
and pumped water from even greater depths
The three steam engines compared

9 How the Coal was Won

How did the miners **work** the coal? Coal seams can vary in thickness from a few centimetres to well over 10 metres. In early mines and into the early years of the 20th century miners used simple tools, such as picks and shovels, to cut the coal. It was not until quite recently that machines have been invented to mine coal.

The main problem was how to stop the roof of the mine from falling in when the coal seam was mined. Most mining accidents are caused by roof collapse—in 1866 when one in every 200 miners died in the pits, half of these were victims of roof fall. A miner ran a fair chance of being buried alive. Wooden posts or **pitprops** were used to support the roof. They had to carry an enormous weight, as a visitor to a mine in the 19th century saw. 'You would think when you were down there you were in a large wood with the props standing about—thousands of them. You would hear them burst like the crack of a whip, but there were

An Anderton Shearer

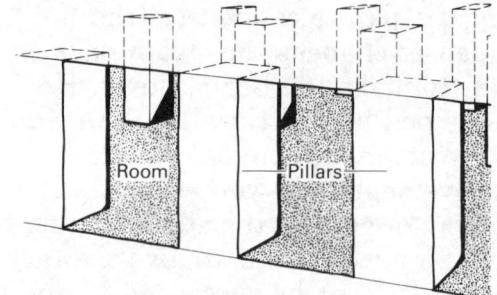

The 'Room and Pillar' mine

Longwall mining

always men there to put in fresh ones at once.'

There are two main ways of supporting the mine roof.

The 'Room and Pillar' mine was the first answer. A bell pit was dug and squared off to form a 'room'. Another 'room' was dug out next to it and so on. The roof was kept up simply by leaving standing pillars of coal. This method was wasteful and is rarely used today.

Longwall mining has been the most popular method for over 300 years. A shaft is sunk and a team of miners work side by side cutting the coal and placing wooden supports behind them as they go forward. The space behind them, or **goaf** is filled in with stones leaving open only a single road to the shaft. Today nearly all pits have mechanical cutters, like the Anderton Shearer. Self-advancing hydraulic supports hold up the roof.

'Mashed up at 40'

What was it like to be a coal miner in the early 1800's? The work was very hard and dangerous. Miners had to cut the coal with hand tools and often had to work in cramped positions for long hours—sometimes on their backs or knees and in damp and dusty conditions. They often worked between 12 and 18 hours at a time. Then there were the dangers—fires, floods, explosions, and roof falls. Accidents were common but did not always make the headlines. Here are two examples:

1854: 7 boys and men fell to their deaths down a 75 metre shaft when one of the chains pulling them up to the surface snapped.

1848: John Guest, aged 18, was killed when he slipped and fell into the **winding** engine. A witness found his 'mangled body' in the gearing.

Also there was the fear of being sacked or thrown out of work for a while, or suffering a cut in wages. Some miners got part of their wages in tickets which could only be used in the mine company's 'truck' shop. Goods in these shops were usually more expensive but often there was no other shop which the miners could use.

A miner's life was miserable in many ways. A hundred years ago miners said: 'For every ton of coal a pint of blood is spilt.' They could expect to live ten years less than men in other jobs. Miners said that they were lucky if they were not 'mashed up at 40.'

But it was not all bad. In 1832 he could earn about £1–20 a week, while a factory worker earned only 35–50p a week. He also got free coal and sometimes his rent was paid by the mine-owner. Some mine-owners even provided doctors for the miners.

A miner hewing coal by by hand

Miners being lowered down the pit by chain and basket

Things to do

1 Make a series of drawings like a strip cartoon to show how a miner's life was hard and dangerous in the early 1800s.
2 Describe what a truck shop was and why a miner would use it.

Young miners

10 Hauling the Coal

A woman carrying a loaded corf

Once the coal has been mined from the rock it has to be moved out of the mine before it can be sold. This means carrying the coal from the face along the mine to the bottom of the shaft—it is usually known as hauling the coal.

In the first drift mines with no shaft it was fairly easy to move the coal out along the drift. It was carried out in wicker baskets called corves which could hold up to 25 kg of coal and were often carried by women and children. Later, the loaded corf might have been dragged along on a rough wooden sled or tram.

In 1841 Patience Kershaw was aged 17. Her job was to drag a corf for 12 hours a day. She described her work as follows: 'I go to the pit at five in the morning. I take my dinner with me, a cake, and eat as I go; I do not stop or rest any time. The bald place on my head is made by thrusting (pushing) the corves. I push the corves a mile or more and back. I am the only girl in the pit; there are 20 boys and 15 men; all the men are naked.'

Young boys and pit ponies

Ponies were also used to haul coal—often tended by very young children. Even today in one or two small pits the ponies are still at work.

As the underground roads went further away from the shaft it became more and more difficult to pull trams and haul corves over the rough ground, often in cramped and damp conditions.

Some mines began to replace the wooden wheels with cast iron ones and to lay iron rails or plates over wooden planks to make hauling easier.

A horse-drawn waggon-way at a drift mine

The first all-iron rails used in hauling coal in mines were laid in 1767. The use of iron helped a great deal but it was still the job of women and children to haul the coal. In the drift mines the rails could often be continued out of the mines to a river or canal where the coal would be loaded on to boats or barges.

The use of steam power by James Watt in the 1780s meant that steam engines could be used underground for pulling tubs of coal. But it was difficult to get large amounts of steam to the depths required and there was always the danger of starting underground fires. It was not until after 1850 that a better kind of power was used—compressed air. This is still used today. So too, is electricity which was first used for haulage in 1883.

Diesel and electric locomotives are used in many modern pits for hauling coal and carrying miners. In earlier times there are reports of children being killed or injured by runaway trams and even today there are occasional accidents in the underground roadways.

Doncaster Evening Post

TUESDAY, NOVEMBER 13, 1973

36 HURT IN YORKS PIT TRAIN CRASH

Thirty-six miners were injured—one seriously—when an underground paddy mail train was derailed at Hickleton Main Colliery, near Thurnscoe, today.

The train, carrying 80 miners, left the rails about 250 yards from the pit bottom.

Things to do

1 Make a picture chart to show how the haulage of coal from the coal face to the bottom of the pit shaft has been improved from the early days.
2 Imagine that you are employed in a mine in 1840 to drag corves of coal to the shaft bottom. Write an entry in your diary for one day describing how you have spent your day.
3 Write a brief account of the work of pit ponies.

11 Winding

Girls carrying coal up the shaft by ladder

The first miners carried their coal up ladders set against the side of the shaft.

By fixing a **windlass** over the mouth of the pit, the coal and miners could be wound up and down the shaft by using a bucket and rope rather like a water well. This could be very dangerous. Here is a typical accident reported in 1842:

> 'In getting off you are at the mercy of the winder. The winder grabs your hand and brings you to land. The unfortunate case of David Pellett who was drawn over the roller by his own uncle and grandfather, just when their attention was drawn to a passing funeral, is a painful example of their unsafety.'

The first improvement on this method was the 'cog and rung **gin**.' You can see how the horses drove the cog to wind the rope up the shaft. The big problem was that the pit mouth was cluttered up with machinery. There was a great improvement when the machinery was set aside from the pit mouth with the invention of the 'Whim Gin'.

A collision of tubs in the shaft

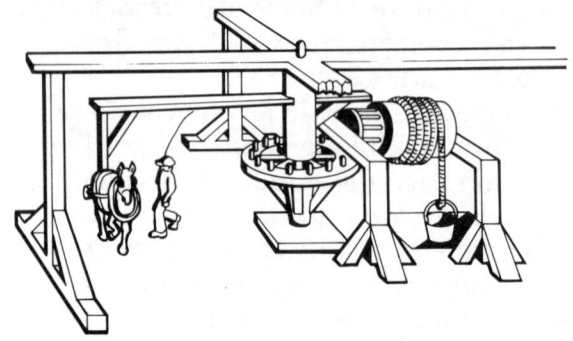

Left: a 'whim gin'; *above:* a 'cog and rung gin'

As the pits went deeper between 1800 and 1850 the use of iron helped to improve winding in the mines. Iron was used to line the sides of the shafts instead of wood. Iron guide rails along the side of the shaft helped to keep the tubs of coal steady, and iron **cages** began to be used to carry the wooden corves to the surface. An important breakthrough was made in 1829 with the invention of wire ropes which made winding more reliable and safe.

The problem of how to get large amounts of coal from the bottom to the top of the pit shafts was not really solved until steam power became available. The first steam winding engine was used in 1784 and was so successful that by 1800 almost all the large coal mines were using them. By 1859, steam winding could raise 800 tonnes each day from a depth of 500 metres—a great improvement on the 100 tonnes a day from 140 metres of the whim gins. Speed and safety have improved enormously over the past hundred years but the newspaper cutting on this page (*Doncaster Evening Post*) reminds us of the dangers of working in the mines.

Raising coal by using a steam engine

13 MINERS DIE IN PIT CAGE PLUNGE HORROR

Thirteen miners died today when a mine cage plunged down a shaft and smashed into the pit bottom.

The accident happened at Markham Colliery, near Chesterfield, Derbyshire, as the cage was carrying 28 men down.

Sixteen men are believed to have serious injuries.

The cage was reported to have over-run its stopping point and crashed into the shaft bottom about 1800 feet from the surface.

One of the first on the scene was 42 year old face worker Mr Matthew Burton, of Staveley.

"It was a terrible scene," he said. "The men in the double decker cage were lying all on top of each other in a heap groaning. I heard one of the men say, Oh, my legs … they hurt."

Rescue workers had to cut away the pit cage gates before they could get at the trapped victims.

Mr Ike Carter, NUM branch secretary, thought the cage went out of control about half way down the shaft.

"The impact must have been terrific judging by the injuries of many of the men, who in some cases were almost unrecognisable."

"The most which can be said is that the cage must have dropped like a brick in the last stages of its descent."

Things to do

1 Make a strip cartoon of pictures to show how the winding of coal has been improved from the earliest times to today.
2 Make a drawing of a whim gin. Write down how the gin works. Why was it an improvement on earlier methods?
3 Find out what 'gin' means. Can you name any other gins? (Clue: cotton industry.)

12 The Mine Children

Children hauling coal past a trapper

If you visited a coal mine in the early 1800s the sight that would have shocked you most would have been the work done by women and children down the mine. A Royal Commission (a government inquiry) was set up in 1840 to report on the mines. Its report shocked everyone.

The report showed that children worked underground before the age of 7 and many started work at 4. The hours were long. 12 hour **shifts** were usual and sometimes children worked for 18 hours at busy times. One girl of 11 said: 'I go down the mine at two in the morning and come up at two in the same afternoon. I go to bed at six to be ready for work.' Many boys and girls never saw the light of day except on Sundays because it was dark when they went down the pit and dark when they came up. A boy of 10, had 'one leg shorter than the other—he was very healthy before he went down the pit.' Some children stood ankle deep in water all day to work the pumps. Sometimes young children were even employed to wind coal and people up the pit shaft. One boy, distracted by a mouse, let go of the winding handle and four boys he was pulling up the shaft were sent crashing to their deaths. There was much brutality— there were reports of children being strapped, beaten and kicked.

A 'trapper' had to open and close the trap doors to let the coal waggons pass. Trapping was not hard work so young children were used for it. They

Children descending a shaft on a rope

A girl hauling coal

had to sit for 12 hours in the pitch dark, opening and closing the trap doors entirely alone. One girl trapper aged 8 said: 'I am a trapper . . . it does not tire me but I have to trap without a light and I'm scared.' A boy, aged 9, said: 'I have been a trapper in the Gauber pit for 3 years. When my light goes out I smoke my pipe.'

Boys and girls also worked at pushing or pulling the tubs of coal on all fours like animals. The picture above shows a girl in a pit at Halifax. She has a chain running between her legs and a belt round her waist. She

Women carrying coal up the shaft by ladder

had to drag a waggon weighing 250 kg a distance of over 300 metres. She did this for 12 hours a day along tunnels usually only about a metre high.

Many women were also employed in the mines, especially in Scotland. They often had to carry coal up the pit shafts climbing ladders and carrying the coal on their backs. One six year-old was found by the Commissioners to be carrying 50 kg of coal 14 times a day up a ladder. The total height she climbed each day was twice the height of St Paul's Cathedral.

Accidents were common. Women fell down shafts or dropped their coal on those below. Trappers fell asleep and were run over by the heavy coal waggons and many boys and girls were crushed by runaway tubs.

Many people were so shocked by the report that something had to be done. Lord Shaftesbury suggested that a new law should be passed to **reform** the mines. As he described the terrible conditions in which children had to work in the mines, many MPs were moved to tears. Yet strangely, many people were even more shocked by the fact that children of both sexes worked almost naked in the presence of naked adults.

Although many coalmine owners tried to stop the new law, the Coal Mines Act was passed in 1842. It said:
1 No women or children under 10 were to work underground in the mines.
2 No child under 10 was to be employed at all.
3 Boys could not operate engines until they were aged 15.
Later in 1850, government inspectors were appointed to make sure that the new laws were kept.

The new laws were important in improving conditions in the mines but there were some problems. It was difficult for inspectors to check the ages of young people. Also the changes made many mining families worse off than before as they could no longer rely on the wages of women and children. Some women even tried to get work disguised as men. However, the Coal Mines Act of 1842 removed the evils of female and child labour for ever.

Things to do

1 From what you have read on these pages write a story about coal mine children with this as your first line: 'When I was four my father sent me to work down the mine . . . '
2 Find out more about Lord Shaftesbury. Why was he called the 'Children's Friend'?

13   King Coal

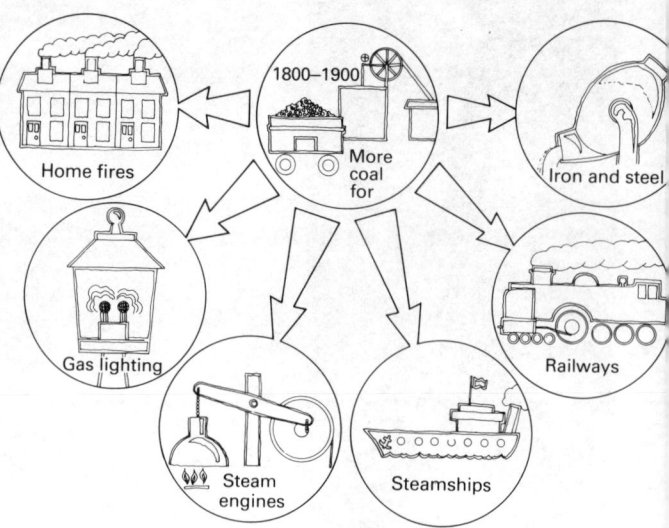

'King Coal' (1800-1900)

About 120 years ago, in Victorian times, coal was in great demand as it was needed for so many things. Britain was the most important industrial nation in the world and British industry relied upon coal. This was the age when coal was King.

At that time large amounts of iron were needed and coal was used to smelt and **forge** iron. Then in the 1850s Henry Bessemer invented his 'Converter' which made cheap steel. This led to a great increase in steel-making and here again coal was needed as fuel for the convertors.

By 1850 Britain had a great network of railways. On these tracks steam locomotives carried people and goods all over the country and the locomotives burned coal as fuel. The railways also helped carry coal to industries where it was needed. So pits no longer had to be next to a river or near the sea. Coalfields far from the sea like Yorkshire became as important as those near the coast.

As steamships gradually replaced sailing ships so more coal was needed for fuel. Coal ports grew up when the ships filled up before they left Britain. Soon coal had to be exported all over the world to coaling stations so that steamships could get coal wherever they went. The coal from South Wales was found to be best for ships.

The rapidly growing use of steam power in the factories was even more important. Steam engines, burning coal for fuel, drove machines in all kinds of industries.

There were many other uses for coal. In the early 1800s William Murdoch found a way to use coal gas for lighting. These were times when Britain's population was growing very quickly. More people means more houses and coal was needed to make bricks. Also each house needed heat and in the great towns it was coal that was burned in the open fires.

Things to do

1 Make a picture diagram to show why much more coal was needed in Victorian times.
2 In 1700 the Northumberland coalfield was the most important in Britain. In 1850 this was no longer true. Write down why this change happened. (Clue: transport.)
3 Find out how coal gas is made. What else can you find out about William Murdoch? (Clue: he worked with James Watt).
4 Find out more about Henry Bessemer and his converter.

14   Deadly Dust

Wherever coal is mined you will find coal dust. This is as great an enemy to the miner as fire or water. The black dust floats in the air and gets in the miners' lungs. This can lead to a terrible disease called silicosis which causes coughing and breathlessness and can lead to early death.

Coal dust can be a killer in a more horrible way as it easily catches fire and is very explosive. A small fire can become a serious explosion by dust carrying it all round the mine. Sometimes it exploded when dynamite was used to blast the coal.

By 1908 it was realized that coal dust can catch fire and explode by itself without a fire starting it.

Today great fans suck out the dust and water is sprayed to keep it down. By laying non-explosive stone dust on all the ledges in a mine, a coal dust explosion can be prevented from spreading and explosions are now rare.

The middle of the 19th century saw some of the worst accidents in mining history. Some of these were explosions probably caused by coal dust. The Oaks Colliery disaster in Barnsley in December 1866 was one.

'The Oaks has exploded', was the cry that winter afternoon. A blast which had been heard several kilometres away had destroyed one cage and had blown another up the shaft into the headgear. There were 340 men and boys down the pit. When rescuers finally reached the pit bottom they found many men dead or badly burned. There were 40 ponies dead with their drivers by their sides. A few survivors were wrapped in cotton wool; some of the dead were taken home.

The following morning, while 27 rescuers were down the pit, a terrific explosion took place shooting up all three shafts at once. All the rescuers were killed. Later there was another great explosion.

Everyone said that nobody could have survived these blasts. But the following morning a signal bell was heard ringing at the bottom of the shaft. A bottle of brandy was lowered down and it returned empty. How could the survivor be reached? All the cages were wrecked. A tub was fastened to a rope which could be lowered down the mine by a small machine. John Mammatt, an official, and a boy called Embleton volunteered to go down. The nightmare journey down took 15 minutes. They could only get one leg inside the tub which spun round like a top. The water pipes had burst and the men were soaked to the skin. At the bottom they found Samuel Brown. He was the only survivor. As they put him into the tub they saw huge fires burning in the distance. All three were hauled to safety, but 361 men had lost their lives. Four years later bodies were still being brought up the pit but for 100 men the Oaks Colliery is still their grave.

Things to do

1 Write a description of the events at the Oaks Colliery during these days in December 1866, as if you were there at the time. You may be an underground worker, a rescuer, a worried bystander or a newspaper reporter.

15 The Long Struggle

A miners' Gala procession in Doncaster

It was impossible for a miner by himself to get better pay and working conditions. If he asked for more money he would probably have been given the sack. But by joining together with other miners and by backing each other up they would be much more powerful. This is the story of how the miners formed their own trade union, which today is called the National Union of Mineworkers.

It was not until 1830 that the first real miners' union was started by Tommy Hepburn in Northumberland and Durham. Within a year the union had won its first strike for better working conditions. At this time the mines were owned by wealthy landowners who opposed the unions, so the next year after 'Hepburn's strike' many coal-owners locked out the miners and refused to give them work unless they agreed to leave the union.

This time the owners were not going to give in. To keep the pits working they brought in workers from other areas. The strike-breaking workers were called 'Blacklegs' and were really hated by the union men as this old song shows,

'Join the union while you may
Don't wait until your dying day
For that might not be far away
You dirty blackleg miners.'

Many union men were thrown out of their homes to make way for the blacklegs. After a violent and bitter struggle the miners were defeated but the owners would only give them work if they promised to take no further part in unions. Tommy Hepburn himself tramped round the pits looking for work and became a beggar. He only got work much later when he promised never to start another union.

During the next 50 years there were several attempts to form unions with little success. Even so, small groups or associations of miners in different parts of the country were working to try to look after the interests of the miners.

In 1889, Ben Pickard, a Yorkshire miners' leader, persuaded miners' associations in several areas to join together to form the 'Miners' Federation of Great Britain'. This new union wanted an 8-hour day for all miners. This was one of their slogans:

'Eight hours work,
Eight hours play,
Eight hours sleep
and eight bob a day.'

Although several areas did not want to join this union at first, it was an important step towards setting up one union for all miners.

Giving soup to the children of evicted miners

In 1893, the Miners' Federation was put to its first real test. The coal-owners tried to cut wages so the union called a strike. The strike was a long and bitter one. There was trouble at Featherstone Colliery in Yorkshire and soldiers were called to keep order. During a demonstration the soldiers fired on a crowd of miners—two were killed and 16 injured. This made many people support the miners and money poured into the union funds. Eventually, after the government had stepped in, the miners won—there was to be no cut in their pay.

After this victory the Miners' Federation grew fast, and when in 1908, the miners of the North-east joined, all miners were in the same union. In that same year, a law was passed giving the miners their 8-hour day.

There were still problems, though. The miners and the coal-owners did not trust each other. Some of the miners' leaders, like A. J. Cook, believed that the miners themselves should run the mines.

In 1914 the miners made an agreement called the 'Triple Industrial Alliance' with the railwaymen's union and the transport workers' union. They agreed to help each other out. The government ran the mines during the First World War (1914-18) to make sure there was enough coal and the miners were well off. But after the war the mines were returned to their owners even though government experts said in an inquiry that the mines should still be run by the government.

The 1920s were not good years for the mines. The pits were old-fashioned and British coal was much dearer than most coal from other countries. To try to cut the cost of British coal the owners cut the miners' wages. The miners asked for help from the Triple Industrial Alliance, but on 'Black Friday', 15 April 1921 their friends let them down. Although they went on strike by themselves the miners soon lost and had to accept the wage cuts.

In 1925, the owners again tried to cut wages. Using the slogan, 'Not a penny off the pay, not a minute on the day', the miners, led by A. J. Cook, again asked for help from their friends. This time, on 'Red Friday' the other unions agreed to help. After a government inquiry had said that the miners must take the pay cuts, the miners asked for help from all the unions in the country and got it. On 3 May 1926 a General Strike began.

Things to do

1 Make a time-line of the history of miners unions of what happened on these dates: 1830, 1832, 1889, 1893, 1908, 1921, 1925, 1926.

The 'Triple Industrial Alliance'

16 Defeat, Despair and Hope

1914: the Triple Alliance

1914–18: Government run mines to keep up coal supplies

1919: Mines given back to owners who cut miner's wages

1921: Miners appeal to allies for help without success

1925: Mines in more trouble, owners cut wages again

1925: Miners appeal for help from other unions and succeed!

TUC calls General Strike to help miners

General Strike called off. Miners stay out and gain nothing

On 3 May 1926, the country almost came to a standstill as the **General Strike** called by the Trades Union Congress began in support of the miners. It lasted only 9 days. The other unions went back to work leaving the miners to fight alone. They stayed out on strike for six months, but were finally forced back to work as their union funds ran out. They suffered six long months for nothing and had to accept wage cuts and longer hours. The miners were never to forget the General Strike.

By the 1920s and 1930s Britain's King Coal was king no more. Britain had once led the world in coal mining but now its pits were old-fashioned. Other countries had newer and better pits.

In the 1930s the coal industry was in serious trouble as other industries used much less coal. Many mines lost a great deal of money. Miners' wages were cut and many lost their jobs to join the **dole** queue; men were sometimes out of work for up to 7 years.

When the Second World War broke out in 1939 there was a short improvement as coal was again needed urgently. Men were given the chance of volunteering for the mines instead of the forces.

When the war was over, few people wanted to go back to the bad old days. The mines were still out of date and even more run down than before and the miners still did not trust the hated coal-owners.

This was the chance for a new start. The Labour government, elected after the war in 1945, decided to nationalize the mines. This meant that the government would take over the industry to run it for the people of the country. On 1 January 1947, the National Coal Board took over the mines from the owners who were given generous compensation.

How did nationalization affect the coal industry? First, the miners welcomed it. Millions of pounds were spent in bringing the pits up to date. Large new pits were sunk and modern machinery installed. Many small pits which made losses were closed down. Miners' wages and working conditions were improved. There were less accidents and up to the early 1960s coal production increased to provide Britain with the coal she needed.

Things to do

1 Draw a poster as if for display in 1947 showing the advantages of nationalizing the mines.
2 What other industries have been nationalized? Are these industries successful?

Some of the effects of nationalization

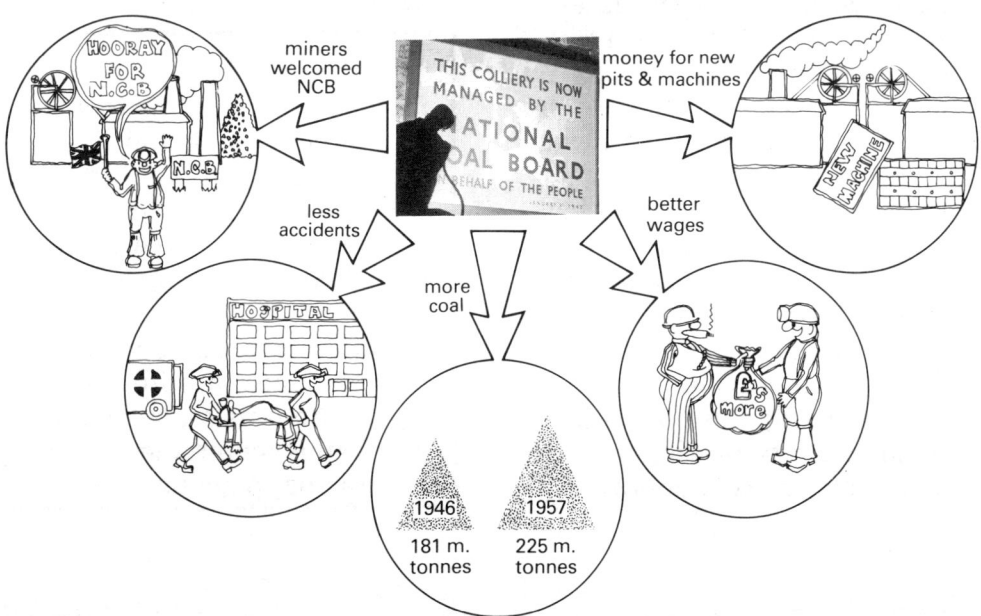

17 Work and Welfare

A coal miner at work in a wet seam

A miner of the 1840s would not have found things very different if he returned to the coal face in 1920. Most miners still worked with a pick and shovel. Naked, or nearly naked, they crouched or lay along the narrow seams clawing at the coal, as the best seams had been worked out. The work often left them stooped and bowed. Accidents still happened and the worst one ever, at Senghenydd in Wales when 439 men died, was in 1913. There had, of course, been changes but the great improvements in automated coal-cutting, power haulage and real safety came only after nationalization.

In many pit villages life centred around the pit as these communities were often cut off from other towns. Inside the terraced pit houses daily life was still geared to the mine as often father and sons worked on shift at the pit. Wives prepared the food for the men to take on the next shift; they warmed their work clothes and later got the tin bath ready for when the grimy men returned home.

Gradually improvements have been made in miners' social lives. Pit-head baths have been built at all mines 'so men could leave their dirt where they got it—at the pit'. Better houses, social centres called Welfare Institutes, and mining colleges have been built. Modern transport has brought the pit villages closer to other communities. Even so, the men still cling to their traditional community pastimes—leek growing in the North-east, choral

Bathing at home after a day in the mines

singing in Wales, brass bands in Yorkshire and pigeon-racing everywhere.

Things to do

1 Draw six diagrams to show how miners' work and lives have changed this century.
2 Write a letter, as if from a miner's wife to a friend in 1900, describing a day in her life.
3 Find out more about popular pastimes in mining communities, such as: brass bands, choirs, leeks, whippet racing, pigeons.

18 A Modern Coal Mine

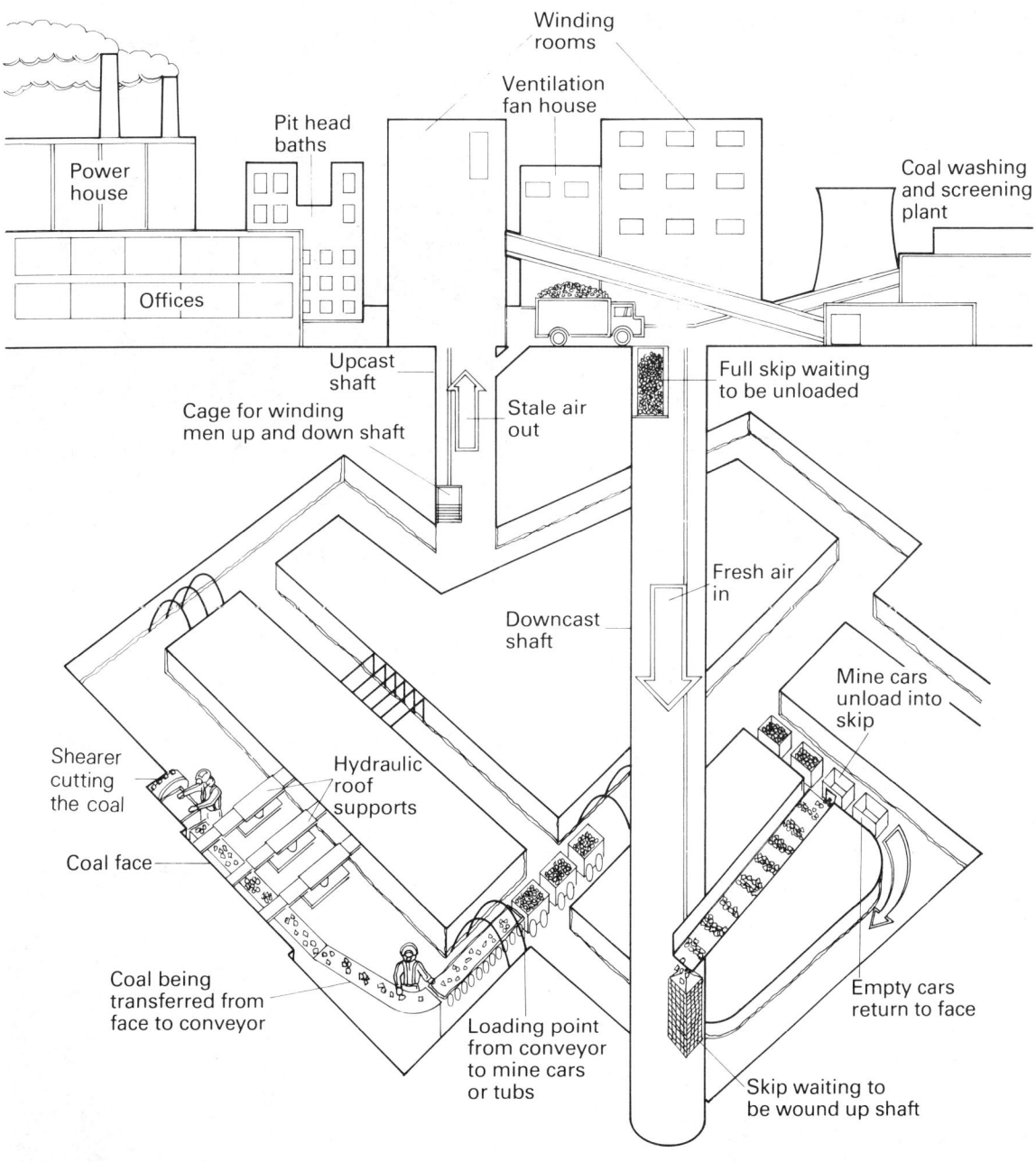

Winding rooms

Ventilation fan house

Pit head baths

Power house

Offices

Coal washing and screening plant

Upcast shaft

Cage for winding men up and down shaft

Stale air out

Full skip waiting to be unloaded

Downcast shaft

Fresh air in

Mine cars unload into skip

Shearer cutting the coal

Hydraulic roof supports

Coal face

Coal being transferred from face to conveyor

Loading point from conveyor to mine cars or tubs

Empty cars return to face

Skip waiting to be wound up shaft

19   The Future

What will happen in the coal industry in the future?

In the 1960s, oil was so cheap and there seemed to be plenty of it that many mines were closed down as it became more and more expensive to mine the coal. There were no plans to start any new mines and coal was becoming less important as a fuel as more oil was used in industry and the home.

But, headlines like this in the 1970s told of big increases in the price of oil.

October 1973

HUGE RISES IN PRICE OF OIL

Many people also began to realize that supplies of oil were being used up very quickly. Unless huge new oil fields are discovered all the oil in the world will be used soon after the year 2000.

In Britain, though, there is enough coal to last about 300 years if we use it carefully. Coal has become important again.

The coal industry is expanding and modernizing now to supply us with enough coal for the future.

The National Coal Board now provides over 200 000 people with jobs.

Almost all the coal is mined today by machines like the Anderton Shearer (*see page 14*).

Every day 190 mines all over the country produce over 280 000 tonnes of coal.

The largest conveyor belt in the world - it is over 8 km long

Plans have been made to open up to 30 new mines by the year 2000. The most important of these is at Selby in Yorkshire. The Selby mine will be the largest single mine in the world with 5 separate shafts, several kilometres apart. It will employ over 4 000 men who will mine over 10 000 000 tonnes of coal a year. Coal will be an important fuel for many years to come.

An NCB poster

People will always need coal

This diagram shows you how coal is used today

Home Heating 10%

Industry 13%

17%

To make Iron and Steel

To make electricity 60%

20 'The Price of Coal'

Aberfan—after the disaster

Although many of the worst dangers of working in the mines have now gone, people who live in mining communities know that their families can still be in peril. It was a terrifying and almost unbelievable disaster that shook not only the small Welsh mining village of Aberfan but the whole world on 21 October 1966.

It was 9.10 a.m. that misty Friday morning. Lessons were about to begin at the Pantglas Junior School. Above the village, on the mountainside, Gwyn Brown and David Evans were working on the local **slag heap** which towered above the village. They were startled by a sound 'like thunder' and watched in horror as the tip began to crumble and slip down the mountain getting faster as it moved. Within seconds a huge wall of black slurry was hurtling down the mountain side towards the school. Howard Rees, a pupil at the senior school, saw:

> 'a big wave of muck higher than a house with boulders, trees, sheep, bricks and slurry in it. It was moving fast, fast as a car goes in a town and was making a rumbling noise like an old train.'

The muck hurled itself against Pantglas Junior School. The children in three classrooms had no chance. They were buried alive as the slurry went on through the school knocking down houses in the streets beyond.

Within minutes a desperate rescue was started, but it was useless. No one was brought out alive after 11 o'clock. Most of the school had disappeared; mothers gathered round the school steps, some weeping, some silent, some shaking their heads in disbelief.

144 people had died—including 116 children, mostly aged between 8 and 10. Two trenches, each 30 metres long, were dug to receive the coffins. A cross 40 metres high was made with the flowers sent from all over the world.

Why did it happen? There had been warnings that a disaster might happen. In 1944 part of the tip slid down the mountain. In 1960 people in Aberfan complained to the NCB of 'filthy, black water' from the tip flooding their homes. Nothing was done. The inquiry into the disaster blamed the NCB for tipping slag on top of streams on the mountain.

A young girl wrote this poem about Aberfan, but it is also a comment on the whole history of mining coal.

> 'Twelve and six a hundredweight,'
>   I heard a neighbour say,
> 'Coal's twelve and six a hundredweight,
>   A terrible price to pay.'
>
> 'It only costs you money,'
>   Said a second wife,
> 'But what of those at Aberfan?
>   The price they paid was life...'

Glossary

Adit: a level passage cut into a hillside to carry away water.

Cage: an iron structure used as a lift in a mine shaft.

Coal face: the area of the mine facing the coal where the miners work to take out the coal.

Coal field: the area where coal is to be found underground.

Coke: a solid substance of nearly pure carbon obtained from coal.

Collier: a coal carrying ship. The word also means a coal-miner, or a sailor on a coal ship.

Dole: unemployment pay.

Drift mine: a level or gently sloping tunnel leading from the surface to the underground coal.

Forge: to shape iron by heating and hammering.

Gallery: an underground tunnel or passage-way in a coal mine.

General Strike: a strike by all workers in the country.

Gin: a shortened word for engine. It usually refers to an engine used for winding.

Goaf: an empty coal seam from which all the coal has been taken. It is filled in with rubble when all the coal has been removed.

Outcrop: the point where the coal comes to the surface.

Pit: a coal mine or colliery.

Pit-props: posts to hold up the roof of the mine.

Pocket (of gas or air): an area down a mine where an amount of gas or air has collected.

Reform: to improve things or remove things that are wrong.

Seam of coal: a section of rock underground which is coal. Coal seams can run for many miles.

Shaft: a vertical tunnel which leads from the surface to the underground passages or **galleries** of the mine.

Shift: the time a person works. Usually shifts work in relays and one person takes over from another.

Skip: a container in which materials are lowered or raised in the mine shafts.

Slag heap: place where all the mine waste material is tipped.

Smelting: the method by which a metal like iron is obtained from its natural mineral or ore by melting.

Waft: to shake something (like a jacket) to make a draught of air to remove a dangerous gas.

Winding: raising and lowering men and materials up and down the pit shaft.

Windlass: a simple machine used for lifting materials up and down a well or mine shaft using a rope and a bucket.

Work the coal: to cut the coal away from the surrounding rock.

© D Hale & M J Vickers 1979

First published 1979
by Edward Arnold (Publishers) Ltd
41 Bedford Square, London WC1B 3DQ

Reprinted 1983

British Library Cataloguing in Publication Data

Hale, D
    Coal mining.—(People and progress).
    1. Coal mines and mining
    I. Title II. Vickers, M J III. Series
    622'.33   TN802
    ISBN–0–7131–0309–4

Set in 12 on 13pt Century Schoolbook and printed in Great Britain by Unwin Brothers Limited
The Gresham Press, Old Woking, Surrey
A member of the Staples Printing Group

Acknowledgments

The Publishers' thanks are due to the following for permission to reproduce copyright photographs:
From *'Coal Technology for Britain's Future'* (by permission of Macmillan, London and Basingstoke): 30b; *The Guardian*: 2; From *'The Industrial Revolution'* by M. E. Beggs-Humphreys (George Allen and Unwin): 10; Mansell Collection: 15t, 15b, 18t, 18b, 19, 20c, 20b, 21; National Coal Board: 8bc, 8br, 14, 15c, 16t, 28t, 30; National Coal Board (Yorkshire): 24t; Popperfoto: 16c,b, 17, 28b; Press Association: 3; Punch: 25; Radio Times Hulton Picture Library: Title Page, 20t; From *'Views of The Collieries in Northumberland and Durham'*, T. H. Hair (1844): 11; Wakefield MD Libraries, from an original postcard in the possession of Miss M. Bent of Kinsley: 24.